Farid Saadaoui

Water Activity Control of PEMFC Using EMR and Inversion Methodology

Farid Saadaoui

Water Activity Control of PEMFC Using EMR and Inversion Methodology

Water Activity Control of PEM Fuel Cell Using Energetic Macroscopic Representation and Inversion Methodology

Noor Publishing

Imprint

Cover image: www.ingimage.com

Publisher:
Noor Publishing
is a trademark of
Dodo Books Indian Ocean Ltd. and OmniScriptum S.R.L publishing group

120 High Road, East Finchley, London, N2 9ED, United Kingdom
Str. Armeneasca 28/1, office 1, Chisinau MD-2012, Republic of Moldova, Europe
Managing Directors: Ieva Konstantinova, Victoria Ursu
info@omniscriptum.com

Printed at: see last page
ISBN: 978-620-0-04931-5

Water Activity Control of PEM Fuel Cell Using Energetic Macroscopic Representation and Inversion Methodology

Farid SAADAOUI

1) LIS2T Intelligence laboratory Artificial for on-board systems, advanced telecoms and telemedicine. Electrical Engineering Department, University Mustapha Stambouli, Mascara. Algeria.

2) CAOSEE Research Laboratory Control, Analysis and Optimization of Systems Electro- Energetic systems, University of Tahri Mohamed Bechar, Bp417, Bechar, Algeria.

Email: farid.saadaoui@univ-mascara.dz

Email: fsaadaoui1969@gmail.com

Abstract: This paper describes a method for water activity control in a Proton Exchange Membrane fuel cell using a new graphic tool of energy systems called Energy Macroscopic Representation (EMR). This methodology allows the interconnecting models of the fuel cell voltage, power, and Relative Humidity (RH) as a function of operating variables such as current, pressure, temperature, the flow of hydrogen, oxygen, and water. The validation of the simulation results illustrates the influence of relative humidity on the fuel cell behavior. In addition, these variables are adjusted taking into account of relative humidity in order to reach a better performance of the system. For this reason, a control method based on inversion, namely the Maximum Control Structure (MCS) is used to make this energy source more accessible and to avoid the disadvantages of over flow and drying case in order to achieve a suitable relative humidity needed for the implementation of this fuel cell.

Keywords: PEMFC, Modeling and Control, Water Activity, RH, EMR, MCS.

1. Introduction

The objective of this work is to study the energy management of a Proton Exchange Membrane fuel cell (PEMFC), based on water management and its effect inside the core of the fuel cell (FC) as a function of hydrogen and oxygen flow rates. It takes into account the humidification of the membrane and its influence on the proper functioning of this system in electrical energy in order to avoid drying and over flow, which can cause the deterioration of the FC.

Due to the high reliability and efficiency of a PEMFC in low temperature, several research works have been carried out in order to use its physical and electrochemical systems as renewable energy in differently applications (mobile, stationary and residential ... etc.).

The PEMFC also called hydrogen cells can be used at temperatures below 100 °C having a wide power range. Such a choice is made due to its operating flexibility, viscosity, performance, solid electrolyte and quick start. These are good converters of hydrogen and oxygen in electricity. The heart of the PEMFC consists of [1-2]:

a- Anode: an electron emitter.

b- Cathode: an electron receptor.

c- Carrier Membrane (electrolyte) separates the anode and the cathode.

Water transport mechanism in electrolyte membrane FC is complex, based on the backscatter phenomenon, which is defined as a function of the moisture gradient between the gases of the cathode and the anode. Backscattering is hard to quantify because it depends on the inlet and outlet humidity and also on the hydration state of the membrane that leads to the electro-osmosis phenomenon which allows the movement of H^+ ions in the electrolyte. An approximation of the electro-osmotic flow is possible because protons move in the membrane only in hydrated form H_3O^+. The flow of water through the membrane by electro-osmosis is in terms of the produced current and the number of water molecules per H^+ ion [2].

Thus phenomena are likely to occur locally, even after cell design simulations, which take into account the movement of electro-osmotic water and the re-diffusion of water from the cathode to the anode in cell configurations and actual operating conditions [2].

Most of the research done on the membrane is based on the microscopic study of the electrolyte membrane; this article presents a macroscopic study. In this work, it is assumed that the hydrogen used at the anode was always saturated with water, while the ratio of saturated air to dry air could vary for the cathode.

It is interesting and indispensable to improve the FC lifetime and to optimize its performance to keep the rate of relative humidity of the membrane constant.

A fuel cell produces between 0.7 and 0.8 V, and in order to have higher voltages, multiple cells are stacked in series [3-4].

The various components of a PEMFC system are represented in Fig. 1. This model consists of five fields of productions:

1-Thermodynamic domain: it is the main domain of electricity production (the core of the cell (anode, cathode and membrane)).

2-Electrical domain: it is the presence of the stack current, the production of a desired constant voltage *Vcell* to a load.

3-Thermal domain: it refers to heat production in the course of the electrochemical reaction.

4-Water management domain: it is the production of the water flow required for the moistening of the membrane.

5-Fluidic domain:

-At the cathode level: inlet of pure air (oxygen).

-At the level of the anode: hydrogen inlet.

The design of each area is an important part of the one of the complete system; it is a fundamental element in achieving high-performance fuel cell systems [5].

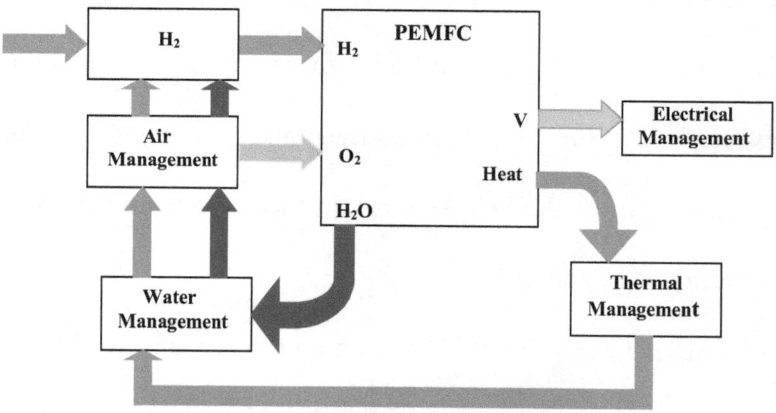

Fig. 1-Synoptic diagram for an operating fuel cell model [6-7]

Generally, the water production in a fuel cell is located at the oxygen side at the cathode. For the anode, no water is produced, but the major problem is that the water produced on the cathode side can migrate through the membrane to the side of the anode where it can accumulate. The majority of the studies are carried out to eliminate the water that can be presented to the anode, this research is based only on the air supply and not on the hydrogen [6].

In our simulation case we consider the temperature as constant as T=363°K. So RH that physically can't exceed 100%, if RH >100%, the exhaust air saturated and unable to evacuate all the water (produced and contained in the incoming air) of the FC. As a result, there is accumulation of water in the system [6].

This indicates that PEMFC can conserve water in the air. Moreover, it is widely known that hot air can contain more water than cold air.

This study concerns the influence of the wet domain on the PEMFC model, with reference to the experimental approach of Rankine that depends on the temperature of the incoming air and has saturation pressure that tends from 0% (dry air) to 100% (humid air) which is the understudy case [6].

From the operational synoptic diagram described in Fig. 1, a model based on EMR methodology of a fuel cell energetic system is developed to assess the influence of input variables such as current, flows rate of (hydrogen, oxygen and water) on the behavior of voltage and relative humidity[6-7].

The electrochemical mathematical model of PEMFC is developed by semi empirical equations in order to obtain simulation results to develop and visualize the concern for the influence of water management on the system.

Therefore, we try to define the specific characteristics of the PEMFC and a control strategy by the MCS obtained through modeling and simulation by considering the different control structures for the exploitation of energy taking into account the inputs and outputs of the system.

The simulation results section is dedicated to the discussion and analysis of the behavior of the system whose goal is to achieve a good design of such a model and lead to a direct chain of conversion of chemical energy from fuel to electrical energy.

2. PEMFC Model Using EMR

The specific model is adapted to develop the control structure of a fuel cell system that must have the characteristics (energy, modular, suitable for multi-complex physical domains of different domains and the synthesis of the MCS).

The reason for choosing EMR for this modeling process is developed for the synthesis of complex systems (electromechanical, electrochemical, fluids mechanics, physical and electrical ...) [8].

In order to facilitate the analysis (structural analysis, etc.) and also the synthesis (from the concept of bi-causality), more recent formalisms have been deduced from Bond Graph such as the Power Oriented Graph or the Power- flow Diagram. These tools are particularly adapted to the design of systems.

Furthermore, when a representation is focused on the functionality of each part of the structure, it is called the functional description. This is the case not only of block diagrams, but also Causal Ordering Graph (COG) and EMR.

These representations are especially adapted to the identification of the control of systems [9-10].

The EMR is an extension of graphical formalisms: the COG; it is a new readable macroscopic representation that is more accessible and easy to use. It was introduced in 2000 by the L2EP Laboratory, at the University of Lille in France. It is a suitable method for the development of energy control systems such as (fuel cells, electromechanical systems, hybrid vehicles, tramways, metros and wind turbines… etc.) [9-10-11].

The EMR is composed of four pictograms :

- ***1-energy sources elements:*** green ovals, terminals of the systems,

- ***2-accumulation elements:*** orange rectangles with diagonal line, energy storage,
- ***3-conversion element without energy accumulation:*** various orange pictograms, energy conversion,
- ***4-coupling elements for energy distribution:*** orange overlapped pictograms, energy distribution [9-10-11].

It is based on the action reaction principle which organizes the system as interconnected subsystems according to the integral causality. What's more, an inversion of this description leads to macro-control blocks. The product of action and reaction variables between two elements leads to the instantaneous power exchanged [9-10-11]. **Table 1**.

Table 1: The basic elements of EMR [9-10-11]

ENERGETIC MACROSCOPIC REPRESENTATION			
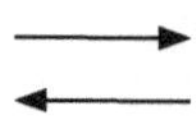	Action and reaction variables	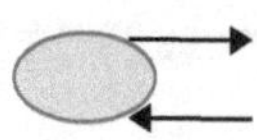	Energy source (system terminals)
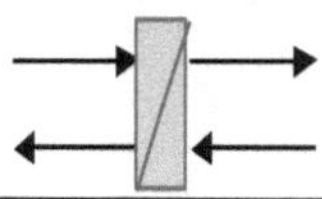	Energy source (system terminals)	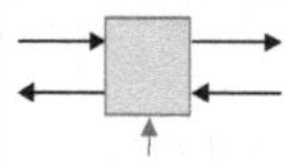	Mono-physical converter (energy conversion)
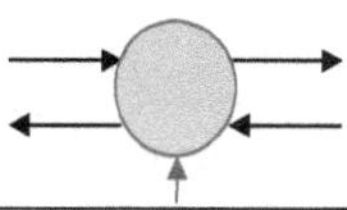	Multi-physical converter (energy conversion)	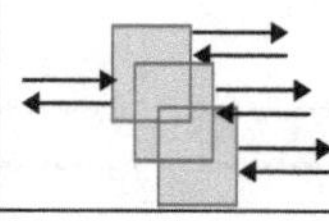	Mono- physical coupling (energy distribution)
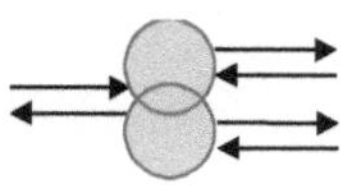	Multi-physical coupling (energy distribution)		

A model fuel cell was proposed by T. Zhou which is a simplified version of that of Hissel [12]. Our proposal consists in adding the constraint of humidification of the membrane by the flow of water introduced into the air, and the influence of relative humidity on the proper functioning of the system, as illustrated in Fig. 2 or the proposed model of the PEMFC stack with different couplings:

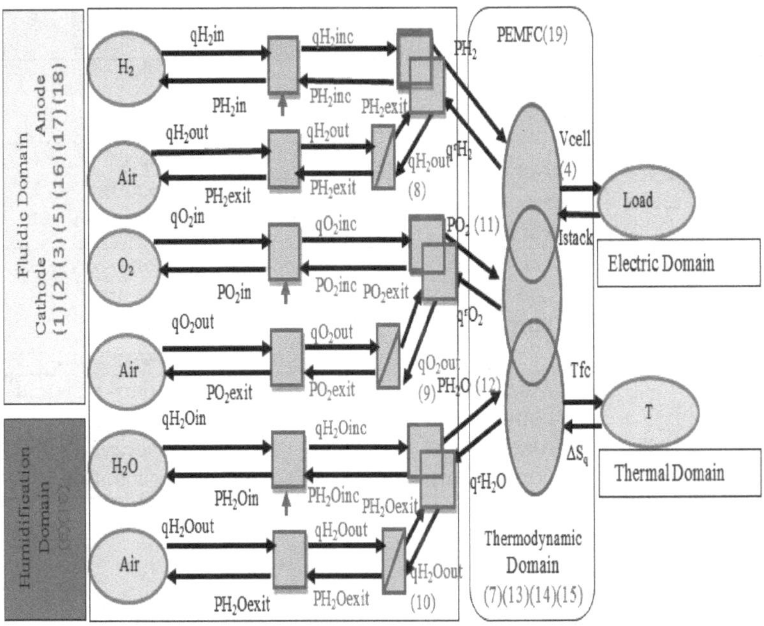

Fig. 2-The proposed PEMFC Representation by EMR

- ***A coupling of the thermodynamic domain with the fluidic domain:*** hydrogen, oxygen and water flow rates were introduced at the anode, the cathode and the membrane.

- ***A coupling of the thermodynamic domain with thermal domain:*** As the temperature variation directly affects the system, a constant temperature has been taken in our case.

- A coupling of the thermodynamic domain with humidification domain: this implies variation of the *RH*, which is our concern.

- A coupling of the thermodynamic domain with electrical domain: by producing a *Vcell* voltage to the load, as shown in Fig. 2.

For this configuration, the equations of the different domains are as follows:

2.1. Thermodynamic Domain

The voltage of a cell given according to the law of ohm:

$$V_{cell} = E_{nerst} - V_{act} - V_{ohm} - V_{con}$$

With E_{Nerst} is the Nernst equation, V_{act} is the activation loss, V_{con} is the concentration loss and V_{ohm} is the Ohmic loss.

The overall tension is as follows [13]:

$$V_{cell} = 1,29 - 0,85.10^{-3}.(T - 298,15) + 4,31.10^{-5}.T.\left[ln(P_{H_2}) + \frac{1}{2} ln(P_{O_2}) \right] -$$
$$\left[\xi_1 + \xi_2.T + \xi_3.T.ln(C_{O_2}) + \xi_4.ln(I_{stack}) \right] - \quad (1)$$
$$B.ln\left(1 - \frac{J}{J\max} \right) - I_{stack}.(\frac{t_m}{\sigma} + Rc)$$

P_{H_2}: is the partial pressures of hydrogen (atm), P_{O_2} : is the partial pressures of oxygen (atm) and T the fuel cell temperature (°K).

$\xi_1 = -0,948$, $\xi_2 = 0,00286 + 0,0002.ln(\mathrm{A}) + \left(4,3.10^{-5}\right).ln(C_{H_2})$

$\xi_3 = 7,6.10^{-5}$ and $\xi_4 = -1,93.10^{-4}$ are experimentally defined parametric coefficients.

$B = \frac{RT}{nF} = 0,016V$: is a parametric coefficient, which depends on the fuel cell and its operation state and *J*: represents the actual current density of the cell (A/cm^2).

J_{max} = 1.5 A/cm^2 is the current density limit.

$Rc = 0.00003$ Ω is the stack internal resistance and t_m is the membrane thickness.

[13]: give the concentration of oxygen Co_2 dissolved in a water film interface in the catalytic surface of the cathode in (mol/cm^3):

$$Co_2 = \frac{Po_2}{5,08.10^6.e^{(-498/T)}} \tag{2}$$

σ_m: Is the specific conductivity of the membrane proton (Ω.cm) which depends on water activity in the membrane, called Nafion conductivity and is obtained experimentally by the following equation [14]:

$$\sigma_m = (0,00519\lambda - 0,00324).\exp\left(1268.(\frac{1}{303} - \frac{1}{T})\right) \tag{3}$$

Alternatively, λ is the water content of the membrane given by the following relation.

$$\lambda = \begin{cases} 0,043 + 17,81.RH - 39,85.(RH)^2 + 36.(\mathrm{RH})^3, & 0 \le RH \le 1 \\ 14 + 1,4.(\mathrm{RH} - 1) & 1 \le RH \le 3 \end{cases} \tag{4}$$

2.2 Fluidic Domain

Both partial pressures of the three inputs gas and water at the PEMFC cell and the defined output rates are by the following electrochemical relationships developed by M. Y. El-Sharkh. The equations of state become:

$$\frac{d}{dt}(P_{H2}) = \frac{RT}{V_{an}}(q_{H_2in} - q_{H_2out} - q^r_{H_2}) \tag{5}$$

$$\frac{d}{dt}(P_{O2}) = \frac{RT}{V_{an}}(q_{O_2in} - q_{O_2out} - q^r_{O_2}) \tag{6}$$

2.3 Humidification Domain

$$\frac{d}{dt}(P_{H_2O}) = \frac{RT}{V_{an}}(q_{H_2Oin} - q_{H_2Oout} - q^r_{H_2O}) \tag{7}$$

P_{H_2inc} , P_{O_2inc}, and P_{H_2Oinc} : the controlled partial pressures of each gas inside the fuel cell (atm) Fig. 2.

V_{an} : is the volume of the anode (l), q_{H_2} and q_{O_2} is the flow rate inlet of each gas and q_{H_2O} is the water inlet flow rate by air (kmol/s).

q_{H_2out} and q_{O_2out} are the outlet flow rates of each gas, and $q_{H_2O_{out}}$ represents water vapor on (kmol/s); q_{H_2in} and q_{O_2in} : the inlet flow rates of hydrogen and oxygen of the cathode and anode; q^r_{H2}, q^r_{O2} , and q^r_{H2O} : usage, production also reaction of the gases and water.

The fuel cell model previously depicted by the EMR is clearly shown in Fig 2.

The input variables are (*Ifc*, P_{H2}, P_{O_2}, P_{H_2O}, I_{stack} *and RH)*, ΔS_q: is the heat flow.

The output variables are *(* $q^r_{H_2}$, $q^r_{O_2}$, $q^r_{H_2O}$, *Vcell, Tfc, and Pcell).*

In the reaction rates of hydrogen, oxygen and water flow rates in the cathode, anode and membrane (electrolyte) of the electrochemical relationships are(kmol/s):

$$q^r_{H_2} = 2q^r_{O_2} = q^r_{H_2O} = \frac{N_0.I_{stack}}{2.F} = 2.K_r.I_{stack} \tag{8}$$

Or N_0 : is the number of cells in the PEMFC stack.

$$q_{H_2out} = K_{H_2} P_{H_2} ; q_{O_2out} = K_{O_2} P_{O_2} ; q_{H_2Oout} = K_{H_2O} P_{H_2O} \quad (9)$$

❖ **Water Activity in PEMFC Formula**

This formula was developed by L. Boulon and all in order to calculate the relative humidity of a fuel cell at the outlet of gases, studying the influence of air and water management on humidification problems.

$$RH = \frac{P_{H_2O}}{P_{sat}(Tair)} \quad (10)$$

P_{H_2O} : The partial pressure of water vapor and

P_{sat} : The saturated vapor pressure.

$$P_{sat} = 10^5 \exp\left(13.7 - \left(\frac{5120}{T_{air} + 273.15}\right)\right) \quad (11)$$

The water content of the outgoing air is analyzed to estimate relative humidity:

$$P_{H_2O} = P_{exit}\left(\frac{0.420 + \lambda_a \psi}{\lambda(1+\psi) + 0.210}\right) \qquad (12)$$

P_{exit} : is the pressure at the output of the stack, for an atmospheric fuel cell λ_a (atmospheric).

λ_a = 2. The air stoichiometry calculated as a function of the molar flow of incoming air and the molar flow of moist air given by the following relation:

$$\psi = \left(\frac{q_{H_2Oin}}{q_{O_2in} + q_{rest}}\right) \qquad (13)$$

$$q_{O_2in} = \frac{\lambda_a . I_{stack}}{4.F} \qquad (14)$$

and $$q_{rest} = 3.76.q_{O_2in} \qquad (15)$$

q_{rest} : is the remaining molar oxygen flux.

Finally, relative humidity is given by the following equation (%):

$$RH = \frac{P_{H_2O}}{P_{sat}(T)} = \frac{P_{exit}\left(\dfrac{0.420 + \lambda_a \psi}{\lambda(1+\psi) + 0.210}\right)}{10^5 \exp\left(13.7 - \left(\dfrac{5120}{T_{air} + 273.15}\right)\right)} \tag{16}$$

ψ: is an adjustable parameter with a possible maximum value of 23.

According to S. Wasterlain , the advantage of a higher cell temperature over the performance of a FC is visualized on the polarization curves. It is shown that for better gas diffusion, the temperature is high since a slight drying of the membrane was detected at a cell temperature T=60°C [15].

3. Model test and validation

The aim of this section is to understand the impact of the operating variables on the behavior of the FC voltage and to improve the performance of the PEMFC model by a proposed regulation. MATLAB/SIMULINK is used to model the system through semi-empirical equations with the parameters cited in **Table 2** and to simulate our 3D figures to analyze the results. These simulation results are as follows:

Table 2: Model Parameters [16-17-18]

PARAMETERS	VALUES
Faradays Constant, F	9684600 C/Kmol
Universal gas Constant, R	8314.47 j/Kmol.K
Number of cells, N_0	100
Hydrogen valve constant, K_{H2}	$4.22*10^{-5}$ Kmol/S.A
Oxygen valve constant, K_{O2}	$2.11*10^{-5}$ Kmol/ S.A
Stack internal resistance, Rc	0.00003 Ω
Constant, K_r	$0.996*10^{-6}$ Kmol / S.A
Surface, A	50.6 cm^2
Membrane thickness, t_m	178 μm
Current stack, Istack	60 A

Therefore, the correct operation of the fuel cell requires a 100% relative humidity [19].

The current demand on the water production (relative humidity rate) in the FC is high; it directly influences the polarization curve, if the quantity of produced water at the cathode is not constant (less or more than *100%*) Figure 5 which, leads to a variation of the *RH* at the FC output according to the power and current density [20].

The performance of the electrolyte membrane varies according to humidity (ratio of dry air to humid air). That is, for intense current densities, an excess of water is produced *(RH>100%)* and this requires reducing the proportion of humid air in order to avoid flooding of the FC. Yet, for low current densities and open potential, water production is reduced and sometimes goes down to zero *(RH<100%)*, which will require an increase in the proportion of humid air to dry air to preserve and maintain *RH* in the membrane constant (which can tend towards *RH=100%*) to avoid drought of the FC. It is the exact determination of these flows that requires control, which is the objective of this paper.

3.1 The Static Characteristics of a PEMFC

a) The current density and oxygen pressure directly affects the FC voltage as shown in Fig. 3. For low oxygen pressures, voltage decreases and for high pressures, it increases rapidly and the voltage pressure characteristic is saturated.

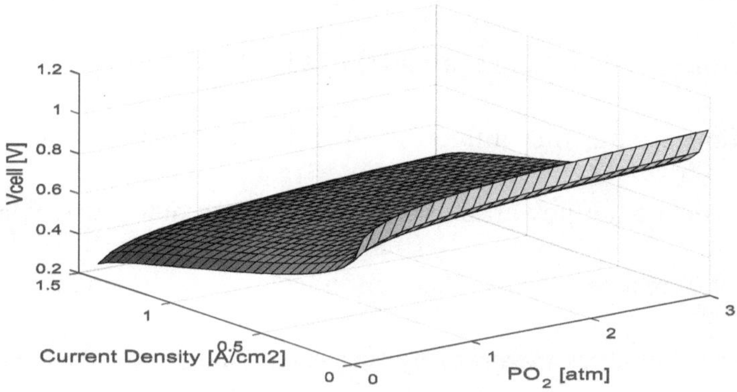

Fig. 3-Variation of PEMFC cell voltage as a function of current density and oxygen pressure with (P_{H2}=P_{H2O}=1(atm),RH=100% and T=90°C).

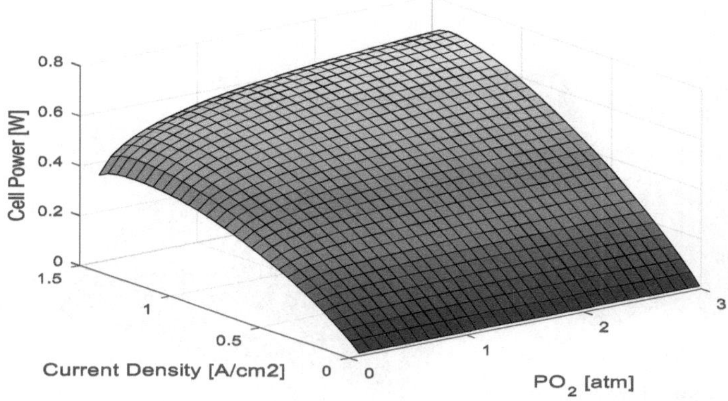

Fig. 4-Variation of PEMFC cell Power as a Function of Current Density and Oxygen Pressure with (P_{H2}=P_{H2O}=1(atm),RH=100% and T=90°C).

b) It is also possible to see in Fig. 4 the range of the power according to these same variables of Fig. 3.

c) Concerning the influence of relative humidity on the behavior of the power shown in Fig. 5, it is noted that when the *RH* reaches *100%* the cell power is at its maximum, relative humidity increases by this value, and power decreases rapidly for high current densities. Therefore, the correct operation of the fuel cell requires a *100%* of relative humidity.

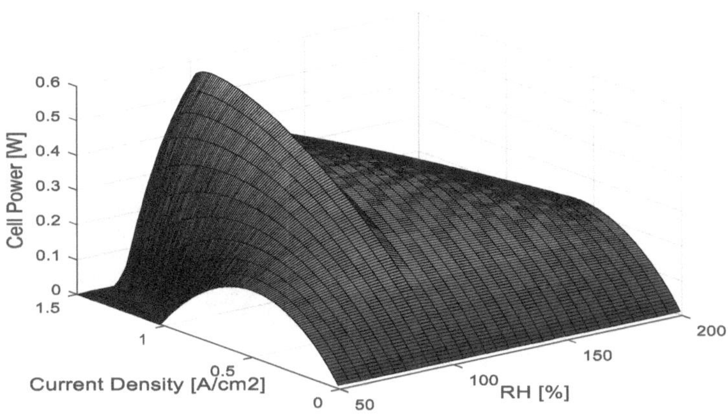

Fig. 5-Variation of cell power as a function of current density and relative humidity with (P_{H2}=P_{O2}=1(atm) and T=90°C).

d) Figure 6 shows the variation of the voltage simultaneously as a function of the current density and of the relative humidity. It is noted that for *RH = 100%* the voltage reaches its maximum and three different parts can be distinguished:

1. Drought part: when *RH* is below *100%.* This is the non-prerequisite part, which can cause membrane defect through the dryness caused by the incoming air.

2. Critical part of flood: *RH* is greater than *100%* (excess water management); in this part, incoming and the producing water is not sufficiently distributed, which causes the flooding of the fuel cell [20].

3. Saturation part: this is the objective of this work, where the flow of water from the inlet (coming from the incoming air and the electrochemical reaction) is equal to that of output, which gives an *RH=100%* and the voltage produced is of course perfectly maximal [21-22-23].

A fuel cell contains two water sources, the wet gas inlet (the steam gas and the hydrated gas with $RH > 0\%$) contain water and internal production.

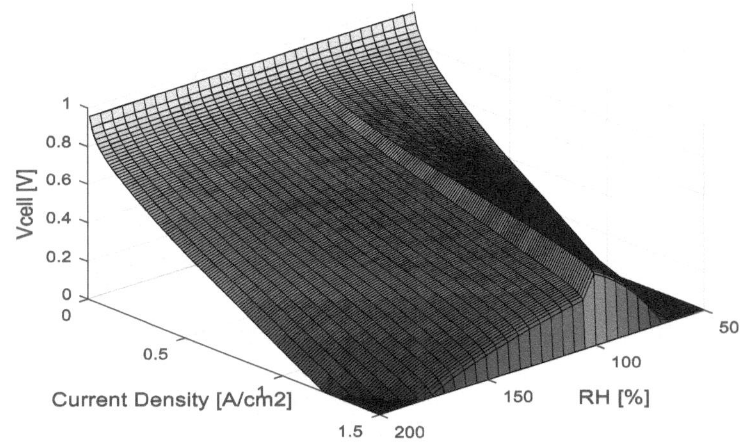

Fig. 6-Variation of cell voltage as a function of current density and relative humidity with ($P_{H2}=P_{O2}=1$(atm) and T=90°C).

e) Figure 7 illustrates the variation of the power as a function of the current density and temperature. In this figure, power increases as a function of the temperature for high current densities.

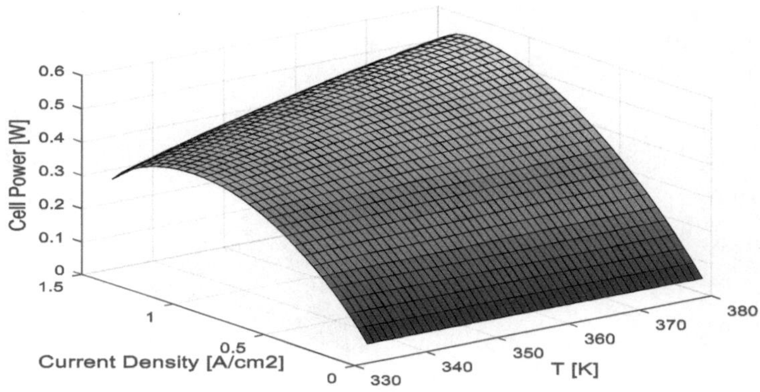

Fig. 7-Variation of cell power as a function of current density and temperature with (P_{H2}= P_{O2}= 1(atm) and RH=100%).

f) Figure 8 shows the variation of the voltage as a function of the current density and the temperature; it decreases when the temperature decreases for high current densities and the vice versa. This characteristic is very sensitive to temperature, which requires the introduction of temperature control.

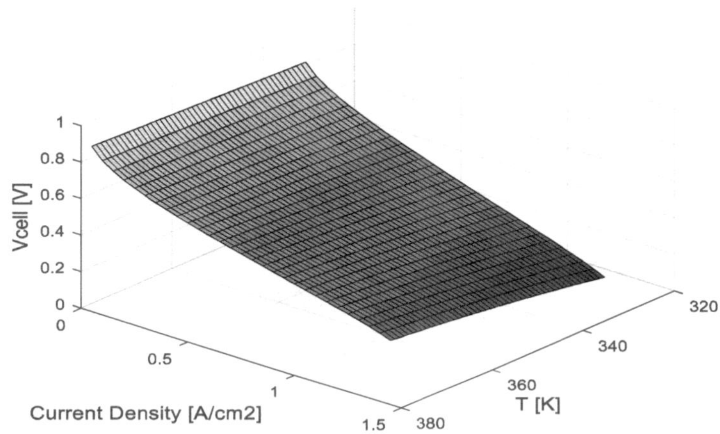

Fig. 8-Cell voltage variation as a function of current density and temperature with (P_{H2}= P_{O2}= 1(atm) and RH=100%).

3.2 The Dynamic Characteristics of a PEMFC

These characteristics make it possible to see the variations of the electrical response (voltage, power and relativity) of the system with various operating parameters (current and incoming pressures of (hydrogen, oxygen and water)) at a constant temperature within *100* seconds.

In practice, the temperature of the PEMFC is regulated to a constant value for active and efficient operation.

Practically, the temperature of the PEMFC is regulated to a constant value for the active and efficient operation.

The input variables are (*Ifc*, q_{H_2} , q_{O_2} , *and* q_{H_2O}).

The output variables are *(Vcell, Pcell, and RH).*

Figure 9 indicates the variations of the voltage cell, power cell, and *RH* as a function of the variations of the different parameters *(Ifc,* q_{O_2in} , q_{H_2Oin}*)* characterizing the uncontrolled PEMFC model. The *Vcell* voltage varies as a function of the current variation of the oxygen, hydrogen, and water flows.

Relative humidity undergoes direct variations according to these same input variables. One can notice that it is between *50* and *150%,* i.e. over *100%,* this encourages us to take into consideration the improvement of this relativity and thus output parameters (*Vcell, Pcell)* and controlled by a well-defined strategy for proper and adequate functioning of this energy system [24-25].

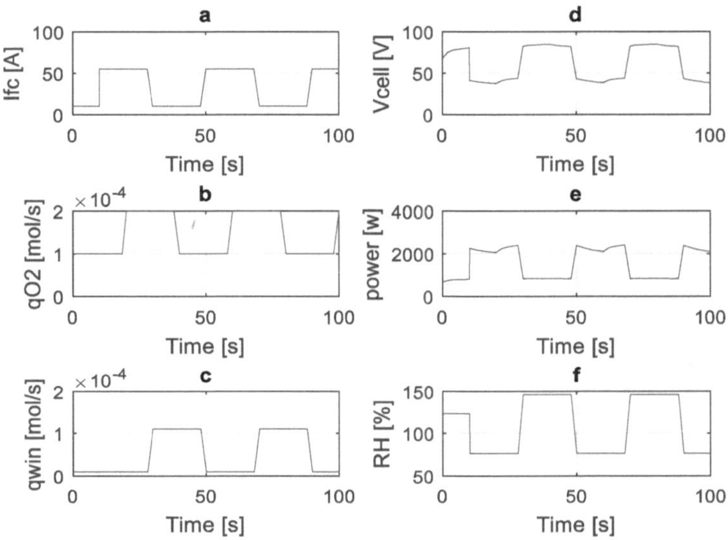

Fig. 9-The dynamic characteristics of a PEMFC

(Ifc, q_{O_2in}, q_{H_2Oin}, Vcell, Pcell and RH).

4. Control strategy by MCS

The MCS structure is successively deducted from the EMR using inversion rules [3] **Table 3**, through the following steps:

1- Defining the control chain (the objective settings variables): in our case, we endeavor to adjust the voltage *Vcell*, power and relative humidity RH Fig. 9.

2- Inverting the adjustment system upward block by block of the control chain.

3- Each block must be separately adjusted.

The main elements used in the maximum control structure are **Table 3**:

- Sources (in green color) that are not inverted.
- Coupling and conversions that are inverted.
- The elements of accumulation which retain their total causalities and are not inverted.

This control methodology consists in controlling the variables (*Ifc*, q_{O_2in}, q_{H_2in}, q_{H_2Oin}) of adjustment system with the objective of controlling the *Vcell* voltage and the *RH*.

4.1. MCS of PEMFC system

The control part designed in this study aims at managing voltage, humidification of the membrane and pressure distribution by acting on input flows. That is why the control strategy given by MCS illustrated in Fig. 10 is used. The power cell and relative humidity are controlled by the incoming flow rates (hydrogen, oxygen and water). The voltage *Vdc* (load voltage) is controlled by a chopper.

Table 3: The basic elements of the MCS [26-27]

MAXIMUM CONTROL STRUCTURE			
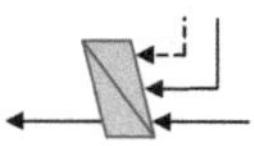	Control block with controller indirect inversion (Closed loop control)		Control block without controller direct inversion (Open loop control)
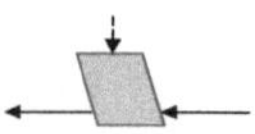	Direct inversion using a disturbance rejection		Strategy (energy management)
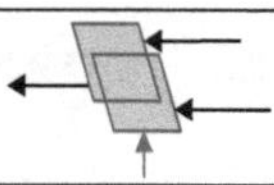	Coupling inversion (weighting)	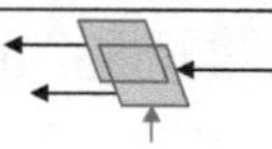	Coupling inversion (distribution)
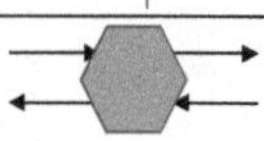	Model or estimator (any pictogram)		Model or estimator

This converter by means of its conversion function is used in the energy conversion of the system in order to control the adjustment chain, making it possible to operate the PEMFC model at a constant reference or predicted reference voltage *(Vref = 60 V)* corresponding to the maximum power point relative humidity reference $RH_{ref} = 100\%$, desired by defining an error ε_3 just like $\varepsilon_3 = RH - RH_{ref} \approx 0$.

4.2. Results and discussions

At the output of the PEMFC model a DC/DC chopper is connected, the FC is assimilated as a controlled voltage source (the voltage of the load is checked) using a PI (Proportional-Integrator) controller this voltage is taken as a reference. The choice of controller has shown its adaptation to the system.

In the physical domain the flows q_{H_2}, q_{O_2} and q_{H_2O} are controlled for our operation.

The coupling control in the MCS part is made in the reaction flow rates: $q^r{}_{H_2}, q^r{}_{O_2}$ and $q^r{}_{H_2O}$ after electrochemical reaction it is compared with reference flow rates $q^r{}_{H_2ref}$, $q^r{}_{O_2ref}$ and $q^r{}_{H_2Oref}$, in such a way as to have errors $\varepsilon_1, \varepsilon_2$ such as $\varepsilon_1 = q_{H_2} - q_{H_2ref} \approx 0$, and $\varepsilon_2 = q_{O_2} - q_{O_2ref} \approx 0$.

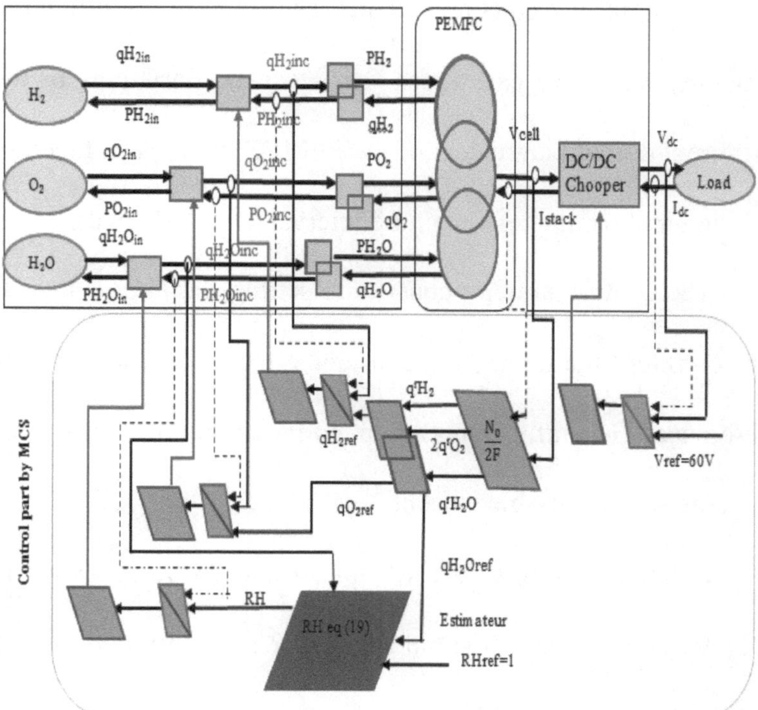

Fig. 10-Proposed Synoptic Control Strategy of PEMFC using MCS

An estimator is appropriate to the control system to estimate relative humidity according to the relationship (16) in Fig. 10. This relativity is controlled by the relative humidity reference $RH_{ref} = 100\%$ needed by defining an error ε_3 such that $\varepsilon_3 = RH - RH_{ref} \approx 0$.

The control results are verified as shown in the following Fig. 11, where we can notice that there are variations of the different output parameters of the PEMFC model as a function of the input parameters in a time interval of *150* seconds.

The control strategy consists in setting the *Vdc* voltage at the output of the model to a constant value of *6OV* intended as reference. Generally, power converters are used to raise this voltage with a variable current.

Relative humidity varies from *50%* to *150%* as shown in Fig. 9 at the output of the model. This *RH* is controlled in such a way that it is around 100% as shown in Fig. 11.

The inversion-based control rests on step-by-step inversion (input and output elements).

This type of control keeps the fuel cell away from flooding and dryness.

Therefore, water management works properly.

Vcell: is the voltage produced by the PEMFC.

Vdc: is the voltage supplied by the DC/DC converter to the payload.

EMR's contribution is designed to design control and energy management systems, compared to other graphical tools, such as Bond Graph or Causal Ordering Graph (COG). The EMR formalism in view of these different characteristics, which reside in the clarity of its structural functional description, its physical concepts, as well as its physical causality, presents a more global real energy view and a deep understanding of its potentialities from a dynamic point of view which focuses on the function of the system and not just on the structure of the system [27].

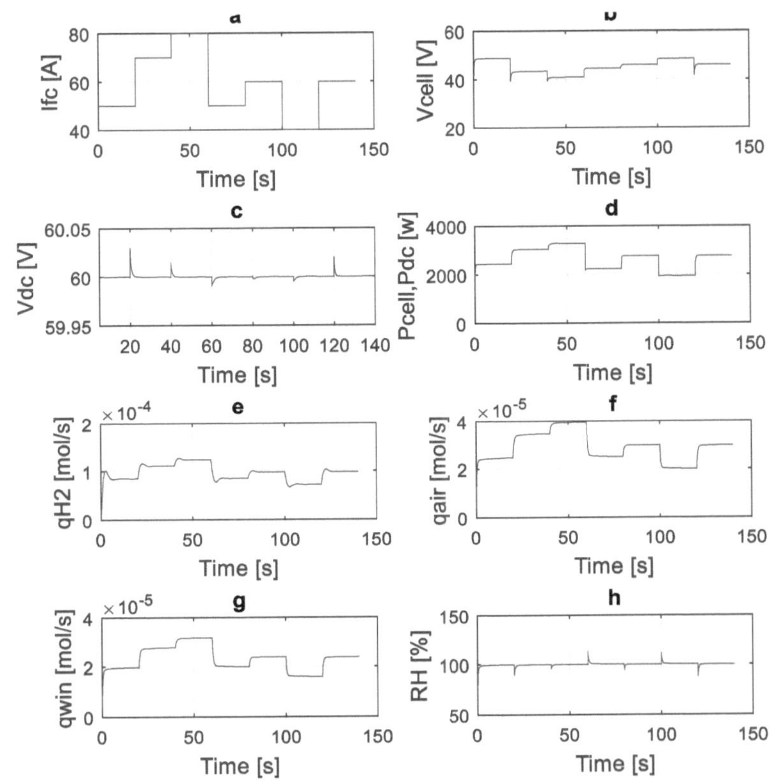

Fig. 11-Control of PEMFC operating variable

5. Conclusion

The influence of water management and control in fuel cell system has been mentioned in this work.

The structure of the PEMFC model is treated as a power source and modeled using the Energetic Macroscopic Representation graph tool.

This method is simple, it analyses the energy exchanges of this energy structure and allows the designer to directly, deduce a simple reliable and relevant control methodology.

In this paper, the Energetic Macroscopic Representation is applied because it seems as the tool the most adapted to the problematic (multi-domain, development of structure command).

The degradation and depletion of water in the PEMFC stack causes a large fall in voltage and can cause deterioration and the end of the system lifespan.

The effect and range of the humidification in the air and the amount of incoming water inside the fuel cell act on the performance of the PEMFC.

To achieve maximum voltage and power, the relative humidity must be around 100%.

The control strategy adopted in this work is the Maximum Control Structure. The inversion of the model consists in adjusting voltage.

Moreover, simulation results show that this energy system can respond easily to the different changes in operating conditions.

Consequently, it is preferable for the supply of electrical energy to design these parameters properly under suitable conditions in order to reduce the voltage losses and to minimize the disadvantages of the fuel cell in order to attain the objective of using the PEMFC with high performance of the cells and to use them in the production of energy.

There has been no electronic estimation device that can help make a relative humidity controller. The authors look forward to creating such a device that allows to estimating, and controlling relative humidity in the PEMFC.

References

[1] J. C. Amphlett, R. F. Mann, B. A. Peppley, P. R. Roberge, and A. Rodrigues: A model predicting transient responses of proton exchange membrane fuel cells, Journal of Power Sources, Vol. 61, No1-2, 1996, pp. 183-188.

[2] O. Warnier, A. Germaing, C. Caldarella and Ch. Charlier Gramme–Liège: Etude des transferts de masse dans une pile à combustible à membrane polymère, Revue Scientifique des ISILF n°22, 2008.

[3] N. Benchouia, A.E. Hadjadj, Modelling and validation of fuel cell PEMFC: Journal of Renewables Energies, Vol. 16 N°2, 2013, pp. 365- 377.

[4] L. Boulon, M-C. Pera, P. Delarue, A. Bouscayrol, and D. Hissel,"Causal Fuel Cell System Model Suitable for Transportation Simulation Applications", ASME Journal of Fuel Cell Science and Technology, Vol. 7, 2010.

[5] K. Mammar, A. Chaker, "Neural Network-Based Modeling of PEM fuel cell and Controller Synthesis of a stand-alone system for residential application", International Journal of Computer Science, Vol. 9, Issue 6, N°1, November 2012 pp.1694-0814.

[6] L. Boulonl, K. Agbossou, D. Hissel, A. Hernandez, A. Bouscayrol, P. Sicard, and M-C. Péra: Energy Management of a Fuel Cell System: Influence of the Air Supply Control on the Water Issues,

International Symposium Industrial Electronics (ISIE), IEEE 4-7 July 2010, pp. 161-166.

[7] R Seyezhai, B.L.Mathur: Modeling and control of PEM fuel cell based hybrid multilevel inverter, International journal of hydrogen energy, Vol 36, 2011, pp. 15029-15043.

[8] L. Boulon, D. Hissel, A. Bouscayrol, and M-C. Pera and P. Delarue: From Modeling to Control of a PEM Fuel Cell Using Energetic Macroscopic Representation, IEEE transactions on Industrial Electronics, Vol. 57, N°6, July 2010, pp. 1882-1891.

[9] D. Hissel, M.C. Pera, A. Bouscayrol, D. Cherenco: Représentation Energétique Macroscopique d'une pile à combustible, Revue Internationale de Génie Electrique, Vol.11, N° 4-5, 2008, pp. 603-623.

[10] A. Bouscayrol, J. P. Hautier, B. Lemaire-Semail: Graphic Formalisms for the Control of Multi-Physical Energetic Systems, Systemic Design Methodologies for Electrical Energy, tome 1, Analysis, Synthesis and Management, Chapter 3, ISTE Willey editions, October 2012, ISBN: 9781848213883.

[11] A. Bouscayrol, B. Davat, B. de Fornel, B. François, J. P. Hautier, F. Meibody-Tabar, M.
Pietrzak-David, "Multimachine Multiconverter System: application for electromechanical drives", European Physics Journal - Applied Physics, vol. 10, no. 2, pp. 131-147, May 2000

[12] T. Zhou: Control and Energy Management of a Hybrid Active Wind Generator including Energy Storage System with Super-

capacitors and Hydrogen technologies for Microgrid Application, PHD, 2009, from the Ecole Centrale of Lille.

[13] Z. Zhi-dan, H. Hai-bo, Z. Xin-jian, C. Guang-yi, R. Yuan: Adaptive maximum power point tracking control of fuel cell power plants, Journal of Power Sources, Science Direct, Vol 176, 2008, pp. 259–269.

[14] N. Yousfi-Steiner, Ph. Moçotéguy, D. Candusso, D. Hissel, A.Hernandez, and A. Aslanides: A review on PEM voltage degradation associated with water management: Impacts, influent factors and characterization, Journal of Power Sources, Vol. 183, Issue 1, 2008, pp. 260-274.

[15] S. Wasterlain, D. Candusso, D. Hissel, F. Harel, P. Bergman, P. Menard, and M. Anwar: Study of temperature, air dew point temperature and reactant flow effects on proton exchange membrane fuel cell performances using electrochemical spectroscopy and voltammetry techniques, Journal of Power Sources, Vol. 195, Issue 4, 2010, pp. 984-993.

[16] K. Mammar, A. Chaker: A Central Composite Face-Centered Design for Parameters Estimation of PEM Fuel Cell Electrochemical Model, Leonardo Journal of Sciences ISSN 1583-0233, Issue 23, July-December 2013, pp. 84-96.

[17] K. Mammar, A. Chaker: Fuzzy Logic Control of Fuel Cell System for Residential Power Generation, Journal of electrical engineering, Vol. 60, N° 6, 2009, pp. 328--334.

[18] G. Naga Srinivasulu, T. Subrahmanyam, V. Dharma Rao: Parametric sensitivity analysis of PEM fuel cell electrochemical

Model, International Journal of Hydrogen Energy, Vol. 36, N° 22, November 2011, pp. 14838-14844.

[19] S. Laribi, K. Mammar and all: Impedance model for diagnosis of water management in fuel cells using artificial neural networks methodology, International Journal of Hydrogen Energy, Vol. 41, No 38, 2016, pp. 17093-17101.

[20] J.G Carton, V. Lawlor, A.G Olabi, C. Hochenauer, G. Zauner: Water droplet accumulation and motion in PEM (Proton Exchange Membrane) fuel cell mini channels, Journal of Energy, Vol. 39, 2012, pp. 63-73.

[21] Y. Chen, OL. Enearu, D. Montalvao and T. Sutharssan: A Review of Computational Fluid Dynamics Simulations on PEFC Performance, Journal of Applied Mechanical Engineering, Vol. 5, N° 6, 2016, DOI: 10.4172/2168-9873.1000241.

[22] K. Agbossou, Y. Dube, N. Hassanaly, K. P. Adzakpa, and J.Ramousse: Experimental Validation of a State Model for PEMFC Auxiliary Control, IEEE trans. on Industry Applications, Vol.45, N° 6, November/December 2009, pp. 2098-2103.

[23] K. Mammar, B. Ould-Bouamama: Analysis of Impedance for Water management in Proton Exchange membrane Fuel Cells using Factorial Design of Experiment (DoE) Methodology, International Mechanical Engineering Congress and Exposition IMECE, Proceedings of the ASME, San Diego, California, USA. November. 15-21, 2013, DO: 10. 1115/ IMECE 2013-63475.

[24] W. Friede, S. Raël and B. Davat, Mathematical model and characterization of the transient behavior of a PEM fuel cell, IEEE

Trans. on Power Electronics, Vol. 19, N° 5, 2004, pp. 1234-1241. DOI: 10.1109/TPEL. 2004. 833449.

[25] J. T. Pukrushpan, A. G. Stefanopoulou, and H. Peng: Control of Fuel Cell Power Systems: Principles, Modeling Analysis and Feedback Design, Advances in Industrial Control Springer UK, 2004. BOOK, British Library Cataloguing in Publication Data, ISBN. 1852338164.

[26] L. Horrein, A. Bouscayrol, Y. Cheng, C. Dumand, G. Colin, Y. Chamaillard, "Influence of the heating system on the fuel consumption of a hybrid electric vehicle", Energy Conversion and Management, Vol. 129, December 2016, pp.250–261.

[27] WWW.emrwebsite.org: ''Modelling and control using Energetic Macroscopic Representation'' Application to hybrid electric vehicles and others, Joint International

Summer School EMR'18, Hanoi University of Science and Technology (Vietnam) and University of Lille (France), 13th – 15th June 2018 Hanoi University of Science and Technology, VIETNAM.

TABLE OF CONTENT

Abstract 1

Introduction 3

PEMFC Model Using EMR 10

Model test and validation 22

Control strategy by MCS 32

Conclusion 40

References 43

FSC
www.fsc.org
MIX
Papier | Fördert gute Waldnutzung
FSC® C083411

Kolibri 360 GmbH
Ferdinand-Jühlke-Straße 7
99095 Erfurt, Deutschland
produktsicherheit@kolibri360.de

Druck:
CPI Druckdienstleistungen GmbH
im Auftrag der
Kolibri 360 GmbH
Ferdinand-Jühlke-Str. 7
99095 Erfurt